MONOPOLY BREAKER

Written by

RON ABRAHAM

authorHOUSE®

AuthorHouse™ UK Ltd.
500 Avebury Boulevard
Central Milton Keynes, MK9 2BE
www.authorhouse.co.uk
Phone: 08001974150

First published by AuthorHouse 3/3/2010

ISBN: 978-1-4490-9193-4 (sc)

This book is printed on acid-free paper.

To those who are struggling to persevere and succeed in the midst of opposition, I dedicate this poem.

A child is born
    equal to others at the start,
    so nurturing has to play its part.

The child grows and is able to take strain,
    along the way unavoidably it endures pain.

Disappointments come along
    bringing hunger with them as their companion.

Friendships are made.

With and without accolade,
    hardships are faced,
    discrimination of class, discrimination of race,
    over-qualified, under-qualified,
    over-experienced, under-experienced.

Returning to the place of birth,
    the child discovers that discrimination
    is as strong as ever.

In the telecom industry,
    working with audacity against the constitution,
    it was claimed to be illegal to apply for permission.

Such arrogance and abuse
    made talking of little use.

Legal was the route that was taken,
    for the monopoly to be broken.

Hidden behind smiles came a master of delusion,
    using lies and innuendos as a weapon.

Professing it to be all in fun,
    he caused the fighter to be brought down.

Uncle Tom thought things would get better,
    but alas he discovered later,
    that he had thrown out the baby with the bath water.

~ RON ABRAHAM

DEDICATION

This book is dedicated to my mother, **Mrs Una Darroux,** and the many Dominicans who fell before breaking through barriers in their respective fields in their homeland. This book is also dedicated to a man whom I consider the greatest lawyer from Dominica, **Senior Counsel Jenner B.M. Armour.**

ACKNOWLEDGMENTS

Thanks to **Christine Bernabe** for inspiring me to write this book and for her assistance. I would also like to express appreciation to **Benoit Esprit, Rena Abraham, Derrick Abraham** and others who helped in the process of seeing this book come to fruition. Thanks also to the **Sun Newspaper** for allowing the use of some articles.

CONTENTS

INTRODUCTION

This book is about a man from humble beginnings; a man who was formed out of hardship and nurtured on discipline; a man who emigrated from his small home country of Dominica to the UK and returned some twenty-six years later to contribute to the progressive development of his beloved island home.

His mother, Mrs Una Darroux, gave him his Dominican roots, and he grew up in an environment of constant needs. This would later be coupled with British grit acquired from many long years spent living in the UK. This was valuable preparation, as it propelled him to overcome many obstacles in life, including one of the toughest obstacles he would face: that of striving to bring television, economically priced quality telephone and Internet access to Dominica by battling against a multinational conglomerate called Cable and Wireless Dominica Ltd (referred to in this book as simply Cable and Wireless).

Life, however, had not prepared him for the nastiness, jealousy and backbiting that would soon follow his victory in the battle against Cable and Wireless.

I am that man from humble beginnings, and this is my "David and Goliath" story. May it encourage you to never give up no matter what difficulties you must face in life.

Chapter 1 – Beginnings

Dominica is a small island of two hundred ninety eight square miles located in the Caribbean between the better-known islands of Guadeloupe and Martinique. It is English speaking and not to be confused with the Dominican Republic, which is a much larger Spanish speaking nation further north in the Caribbean.

I was born of my mother, Una Darroux nee Lestrade, and my father, Andrew Abraham, in Dominica on 21st November 1941 at the Princess Margaret Hospital in Roseau.

Our family name of Abraham came from my great grandfather Joseph Abraham, who travelled to Dominica from Israel and married a Carib lady known as Ma Soosoute. They had a son named Joseph who had three sons, one of whom was named Andrew – my father.

Even before birth, I faced difficulties. I survived an abortive attempt by my mum, which was something she didn't want to do, but she was pressured by her family because they considered her too young to have a baby in her unwedded state. However, she received loving support from one of her sisters, my aunt Mona Lestrade. Thanks to her, my mum was encouraged to keep the baby (me) and I was allowed to be born.

I was weaned naturally after being fed primarily on breast milk, but this formed a bond with my mum that was to last for many decades. When I was a young boy growing up in St Joseph, where I stayed with my grandmother while attending St Joseph Primary School, hardship and hunger were my constant, close companions. I can still remember going to Otrobando regularly to pick mangoes to eat to stave off my incessant hunger pangs.

When I was sent to Roseau to stay with my mum, I was enrolled at Roseau Mixed School under Headteacher Fingal, a very strict disciplinarian noted for correcting his pupils with a very thick strap. However, even at such an early age, I found the opposite sex to be a source of distraction from my studies, and so I registered myself (without my parents' knowledge or consent) into Roseau Boys School under Headteacher Lawrence, who was also a strict disciplinarian.

From there I won two scholarships, one to Dominica Grammar School and the other to Saint Mary's Academy. Although I was quite impressed that the grammar school had a cadet corps, I chose to attend Saint Mary's Academy High School under Principal Brother Crane of the Irish Christian Brothers, yet another strict disciplinarian.

I started school in the second form instead of the first because I was already so advanced in my studies. By the end of the school year, I was considered too bright for the third form and was promoted to the fourth form. At the end of that year I was to go to fifth form, but since fifth was merged into sixth, I was placed in the sixth form – the top class. In 1959, at the end of that year, I sat and passed the Cambridge Overseas School Certificate Exams.

In 1961, I immigrated to the United Kingdom. To those of us in the Colonies, England was considered the Motherland, but on arriving there, I quickly learned that it was a cold heartless society where people seemed to have more love for animals than for their fellow human beings in keeping with the common adage of it being a "dog-eat-dog" world. I found it very difficult learning to eat "dog," but I survived by being a quick learner. An example of this took place at my first job, where I worked as an assistant storeman at a company called Magnatex. My supervisor felt it unnecessary for me to have a tea break, but of course, he was accustomed to having his tea break without fail, by the clock. The denial of my break was an unfair workplace practice and one that I could not withstand, and so I left that company.

I then worked as a toffee Boilerman at Callard and Bowser, the famous toffee makers. The fact that I was the first black person to aspire to that position was a major breakthrough for that company, especially because my pay was very good. Despite this groundbreaking success, I somehow knew that was not where I wanted to work as a career, and so I left that job and went to work at Southall Post Office for a lesser amount of base pay but with lots of overtime pay, which more than compensated for the difference. I worked there for a number of years as a Postman, and then I was promoted to Postman-driver and finally to Postman Higher Grade. Again, the pay was good and the working conditions were very good, but again, I knew that was not where I wanted to be permanently and so I left the job, even though my friends thought I was crazy to do so.

Soon thereafter, I obtained a job at a company called Graviner at Colnbrook in Buckinghamshire. I was a trainee prototype wireman for again less pay, but this field interested me greatly so I enrolled as a part day, part night Electronic/Electrical Technician course Southall Technical College at my own expense. The Company allowed me time off to go to school, but without pay, which meant that instead of earning £30-plus pounds a week, I was only earning £5. I endured this, because at last, I was where I wanted to be in a career field I loved and that fascinated me. I was constantly short of money and had very little time to socialise, but I knew it was a worthwhile sacrifice to achieve my long-term goals.

Since I was married at that time and had two young daughters and a son, I needed to find a job with better pay. I accomplished this when I took a job at Harlow New Town in Essex with a company called Cossor Electronics, which agreed to give me time off three days a week with pay to continue studying. I started out as a prototype wireman and rapidly progressed to Technician, Senior Technician, Junior Engineer, Engineer, Senior Engineer and Project Engineer while being qualified with a full Technological Certificate with endorsements in Automation, Control Systems, and Instrumentation. In that job, I designed electronic systems and test equipment to prove that the company's manufactured products met their specifications. Test equipment had to measure to one order better than the manufactured product; i.e., if the manufactured specification was 10 Volts + or - 0.1v, the test equipment would need to be 10 Volts + or − 0.01 v. Designing the test equipment as an engineer required me to stay abreast of advances in the industry by doing much reading and research.

In order to meet the financial needs of a growing family, I also worked part time in Securicor and worked as a newspaper distributor.

Using my electronic design skills, I designed my own disco equipment and began working part time as a DJ, earning what was considered good money for that time.

I then joined the Institute of Incorporated Engineers as a Corporate Member, later becoming a Fellow Member with designation F.I.E.T and registered with the Engineering Registration Board of UK with designation I. Eng.

CHAPTER 2 – THE BIRTH OF AN IDEA

An important turning point came about in my life in 1975. I visited my native country of Dominica due to the death of my grandmother, whom I loved very much, and while I was there, I was appalled at the poor quality of television reception the people had to endure. They had to use a separate radio for audio (voice) while watching the image on the television screen, and when I saw how troublesome and inconvenient this was for them, the idea was born of bringing good quality television to Dominica.

I returned to England and contacted the government of Dominica straightaway. I offered my services as a skilled engineer free of charge to set up a TV station. The only stipulation was that I be given room and board in exchange for my services, but my offer was refused.

Undaunted by the refusal, I decided to return to the region and took up a short-term contract in Guyana. From there I returned to the UK once more to be with my family, and then in 1979, I went to Dominica to set up an electronics repair facility called Maroni Electronics.

Maroni Electronics was set up to be a modern repair and calibration house; we handled the repairs of all electrical and electronic equipment ranging from a toaster to the radar systems on boats. Out shop was well equipped, and for staff I hired some six school leavers and trained them myself. This proved successful; most of them continued in the field, and some still work with Marpin Telecoms (as the company is now called) to this day.

While in Maroni Electronics, I was approached by a foreign entity to set up a TV station but the Prime minister at the time, Dame Eugenia Charles, refused to have TV ownership in foreign hands and refused to grant permission and by extension the licence.

With her refusal, the idea for quality television reception seemed dead, as

I did not have access to much-needed funds, and I certainly didn't have the funds myself. In the meantime, my knowledge of satellite systems was expanded by reading a large volume of unsolicited literature and magazines that seemed to find their way to me, and I devoured them in my hunger to learn more about this growing industry.

I joined the Full Gospel Business Men's Fellowship International, a fellowship of businessmen worshipping and serving God Most High and His Son JESUS, and by the power of the Holy Spirit I became an executive member and attended regular weekly meetings. After one such meeting on a Friday evening, I was in discussion with Charles Maynard, Brian Alleyne, and Eddie Lambert, fellow executive members, about how I was now able to set up a TV receive-only satellite system but had no funds to do it. They suggested I pray to God Most High and ask Him to give me the meaning and reason for this knowledge that I had acquired.

This I did that very night; I asked the Lord for a clear sign, as did Simeon in the Bible. Simeon had asked God to confirm something by causing the ground to be wet but to be dry on a rug he had left out in the open. When it happened just as he had requested of God, Simeon said to himself, "It was too easy," and then he asked God to reverse it, which He did.

In my case, I just wanted a sign that would leave me with no doubt. The weekend passed, and on Monday morning, as soon as I walked into my office, the phone rang. It was a gentleman named Derick Pinard, a man I barely knew, but he caused my breathing to quicken when he said to me, "Let us open a TV station." Knowing that the lack of finances was a tremendous hurdle we would have to overcome, I asked him if he had the money for such an endeavour, and he said no. I took that to be the sign from God that I'd been seeking, because since neither the gentleman nor I had enough money, **only God could make it work.**

At that moment, in July 1986, the idea for the TV station was conceived. Mr. Pinard and I met and agreed to form the company we would call MARPIN: MAR after Maroni Electronics whence it was born, and PIN after Derick Pinard, the gentleman whom God led to initiate that fateful telephone call to me and who was soon to be my business partner.

We agreed the company would be a limited liability corporation and that private sale of shares would take place.

In order to form the company legally, I loaned Marpin the money from an overdraft facility of his my company Maroni Electronics.

The battle of funding began; all local banks refused to fund the project, with one stating firmly that when Marpin had raised $50,000, the bank would match that by lending us $50,000. On the grounds that Marpin had no "tangible security," I began to personally go round to ask people – "beg" is the more appropriate word! – to invest in my new company. Some people insulted me, especially other business people, who took great delight in reminding me that two well known and respected technicians from a very large competitor called Cable and Wireless Dominica, Ltd., which had said the idea could not work, and that if it could, Cable and Wireless would have done it already. This proved to be a serious setback, but I persisted undaunted because I was confident it would work, and so I continued to sell shares. This carried on for months.

In the meantime, I became very unhappy with the minor role being played by the "PIN" arm of Marpin; with my business partner being a banker, I had expected his involvement in presentations, which was never given.

Mr Pinard played a minor role in the formation of Marpin and had no connections with the company shortly after its formation. I even passed along his shares to his sister.

With great difficulty, $15,000 was raised, which was used to do engineering feasibility studies to establish the mode of transmission. Due to the mountainous terrain and the limitations of the reach of off-air transmissions, thus making many repeaters necessary, cable mode was chosen, and so it was agreed to start with four channels of very good quality television.

The selling of shares continued relentlessly until another $50,000 was raised. Feeling certain that the bank that had given us the promise of matching funds (in the form of loans) would take us seriously now that we had raised this large sum of money, we approached them once again for funding, but once again they turned Marpin down citing the same reasons given previously.

I knew a wealthy businessman by the name of Mr John Keller, an American of some means who had settled in Dominica some years before. He was also a visionary who liked to help people when he saw they were onto a good project that had the potential for success. After seeking confirmation in prayer, I approached him with another director, Edward Lambert, who agreed to take $45,000 of shares based on his knowledge of my character proven business acumen.

I went off to Blonder and Tongue in the US for antenna installation and operation training in readiness for delivery of the satellite receiving facilities in Dominica.

In April 1987, the first Earth station was erected successfully, and members of the public were invited to view the images on a black and white television. A local businessman named Waddy Astaphan was so disappointed that the picture was not in colour that he donated a 21-inch colour television so the full quality of the signal could be appreciated.

In those days, satellite transmissions were unscrambled, making the amount and variety of programming very interesting. This also provided the programmer with much more choice and by extension gave the subscribers a greater variety in program choice as well.

Marpin started sending transmissions from its site, which was bought from Mr Daniel Green as shares in the venture; that site was located at Morne Daniel in Canefield.

Having established the quality of the signals, Marpin then ran coaxial cables south to Roseau, the capital, making it possible for the first paying customer to be connected later that month.

At Marpin, we felt certain that with that major hurdle being overcome, the voice of the critics would finally be silenced and getting lending approval from the banks would be plain sailing. However, this proved not to be the case, as the same "well respected" technicians from Cable and Wireless who had previously said our business idea would not work had circulated a new rumour "that signals could not be amplified over long distances without serious deterioration, and so the quality will only be maintained for about a mile." This negative publicity again proved to be a damaging impediment to the sale of shares in Marpin.

Therefore, we reluctantly had to return to the wealthy businessman, Mr John Keller, to ask for a larger take-up of shares. This he did to the tune of $250,000. Once again, he had come through for us in our hour of need, and he believed in us when not many others did.

On behalf of my company, Marpin, I would like to place on record how much we appreciate Mr John Keller, and were it not for that injection of funds by Mr Keller, the project would have died at that point, and if Marpin had survived at all, it would only have been as a video rental company.

CHAPTER 3 – THE TV ERA IN DOMINICA

A new day had begun; Dominica was now one of the first countries of the former British colonies to have multi-channel good quality television. There was great euphoria as "pictures like glass," as it was described locally, were seen on televisions without the need for a radio to pick up voice for the first time in the history of Dominica. This gave the country a tremendous feeling of well-being, and it gave me a deep sense of satisfaction to see such joy among my fellow citizens.

Demand far outstripped supply along the west coast of the island; meanwhile, even though there was demand for TV on the east coast, that region was not yet in service, causing the need for greater expansion throughout the island.

Marpin was ready to meet this demand, and soon, four channels 8 hours a day became eight channels 24 hours a day … then twelve channels 24 hours a day … then fifty-one channels 24 hours a day. The service was extended by microwave to two areas on the east coast of the island and then to almost all of the east coast.

In the meantime, cable television technology was rapidly developing to find ways of using the entire frequency spectrum as the operating frequencies widened from 300 megahertz to 850 megahertz.

LOCAL PRODUCTION

A faint voice started to be heard as early as 1987 as customers wanted to see themselves on TV. This grew to a crescendo until by early 1989, local television production created a weekly news magazine programme called NEWS FOCUS, which was quickly followed by a live weekly talk programme called WHATABOUT, which sometimes turned into such a ruckus that it could have been named What A Bout!

These two programmes gave Marpin tremendous influence on all aspects of life. Culture became more exciting as, for the first time, performers could see how they looked and sounded; young ladies became more fashion conscious; politicians were now aware that everyone around the country could see them and hear them, thus they could no longer hide the truth from their constituents – all for the low price of $30 a month.

Local merchants began selling TVs in large numbers, and hire purchase was introduced to make purchasing TVs easier, thus causing demand for service to grow.

Other islands started to follow Dominica's example by providing cable television service to their citizens as well, and within a few years, cable TV was the order of the day in the islands, all because Marpin had boldly ventured into the realm of cable TV.

The name Marpin became synonymous with television, so that anyone who had TV was seen as "having Marpin."

ROLE OF CONFERENCES

In the meantime, I became more informed than anyone else in the former British colonies in cable television technology, and I was elected to the Board of Caribbean Broadcasting Union as Vice President of Technology. After its rules were amended to classify cable television as broadcasting, I attended most annual cable television conferences to learn about new technologies including multiplexing voice and Internet over cable for utilising surplus spectrum.

As a founding member and shareholder, I also took a key role in forming a Caribbean Cable TV Co-Op. The purpose of this co-op was to overcome the copyright issues that were becoming more prevalent as satellite providers began scrambling programmes in order to make them more difficult and costly to obtain. As always, we were customer focused: the goal of the co-op was to obtain more programming to satisfy the ever-increasing hunger for more channels.

Voice and Internet

Voice and Internet over cable was attracting more and more interest, and at a conference in 1990, trials were available for evaluation. It was now possible to send voice and Internet over cable by only the addition of head-end equipment and customer boxes.

The rush was on among equipment designers in Israel, China, Japan, Canada, and the US to be the first to develop a reliable set-top box that would transport voice and Internet over cable without static, disruptions, or dropped phone calls.

Wherever these VOIP (voice over Internet protocol) boxes were being demonstrated, I made certain I was present, keeping abreast of the latest technology just like I had been accustomed to doing when I worked as a design engineer in Cossor Electronics years earlier. I travelled extensively during this period in the quest for up to date technology and knowledge of the business.

CHAPTER 4 – BATTLING GOLIATH

It was clear that the day would come soon when TV, Voice and Internet would come from the same network, but my company, Marpin, was only a television company. I knew we had to keep up with this ever-changing industry, and so I investigated the legalities and called a shareholders' meeting to authorise the company to alter its legal documents and change its name to **Marpin Telecoms and Broadcasting**, with the result that in April 1994, this change was finalised.

Observing what was happening in Marpin, the goliath company Cable and Wireless Dominica Ltd. signed a new monopoly agreement with the government tightening the hold on telecoms in Dominica. (Neither my company nor I were aware of how tight this agreement was until later.)

While attending a conference in Zurich, Switzerland for CBU, I discovered that Internet access could be provided via the downstream of cable TV service to customers and upon return, upstream the Internet by telephone to complete the loop. Two or three times a day, an international telephone call was made to "post" and collect the mail.

Marpin entered into an agreement with the service provider in Switzerland, who came to Dominica to set up the necessary equipment. I then approached Cable and Wireless to lease circuits for local lines with 800 numbers, which would then be circuited to the US or Switzerland.

The price we were quoted was out of this world: leased circuits to the US would cost $111,500 EC dollars per month, the equivalent of $41,000 USD or £22,300 GBP (see copy of letter below showing these quoted prices). This was a clear indication of the intent of Cable and Wireless (notated on this letter as Telecommunications of Dominica Limited) to stop my company from expanding.

RA 1

**TELECOMMUNICATIONS
OF DOMINICA LIMITED**

Telecommunications of Dominica Lin
P.O. Box 6
Mercury House
Hanover Street
Roseau
Dominica
Windward Islands

Telephone: 448-1000
International: 809-448-1000
Facsilile: 809-448-1111
Tekex: 8625 DOCAW

March 22, 1996

Mr. Ronald Abraham
The Manager
Marpin Television
Great Malborough Street
ROSEAU

Dear Mr. Abraham

As per our conversation, here are the quotes I promised.

Toll Free 800

Toll Free 800 is ideal for increasing responses to advertising, promotions and sales literature, we can facilitate businesses in Dominica with 1-800 service inbound Dominica locally , from most of the Caribbean islands, United States of America and from Canada.

The rates associated are as follows:

Local

A minimum monthly charge of $63.00 for first 300 units and $0.21 per unit thereafter. A unit is three (3) minutes or part thereof.

120,000
= $8,000

Regional	Up to 10 hrs/Month EC $ per hr.	Succeeding Hrs EC $ per hr.
Band 1 (includes Anguilla, Barbados, BVI, Grenada, St. Kitts & Nevis, Jamaica, St. Vincent & the Grenadines and Trinidad & Tobago	114.00	90.00

Page 1

16

Band 3 (includes Antigua Montserrat and St. Lucia)	81.00	65.00
Band 4 (includes Bahamas, Belize Cayman and Turks & Caicos Is.)	190.00	150.00

Please note that there is a minimum monthly charge of one (1) hour, irrespective of usage. Thus in the case of Antigua, for instance, the monthly charge will always be at least $81.00. However, if more than one (1) hour is used, the monthly charges would be the overall monthly consumption times the rate $1.35 (per minute rate).

Leased Circuit

	Installation	Monthly Rental
Leased Line (to USA)	$1,395.00	$111,500.00
⇒ Leased Line (Switzerland)		

⇒ This will be communicated to you since we cannot furnish you with these rates at present.

Please do not hesitate to call if you require further clarification.

Regards

IAN BLANCHARD
MANAGER SALES & MARKETING (Ag.)

Two months later, the matter was pursued again with Cable and Wireless, and a quotation was given for half-circuit (128 kbs) to Switzerland for $30,880 per month, the equivalent of $11,365 USD or £6,176 GBP for what was nothing more than a 0.08 T1 circuit. Please see the letter shown below as evidence of these quoted prices.

Impact of Monopoly Agreement

Although the 1995 monopoly agreement was supposed to be in the public domain, it was very difficult to obtain. However, when it *was* eventually obtained, it was discovered that it was an offence for one to apply for a telecom licence; it was an offence for the minister, an elected official, to receive an application; and it was an offence to grant a licence.

The Prime Minister in 1995, who was well respected at the time, signed this into law, and it was to last for the duration of twenty-five years.

Marpin considered it to be the highest act of colonialism to satisfy an English company's interest in the colonies, but it was like waving a red rag in front of a bull to make it illegal for a citizen of Dominica to apply for a licence without also taking the risk of going to jail, which was how that law was distinctly worded.

Suddenly it became easy to understand why the government could not grant a full telecom licence to Marpin, and why they could not force Cable and Wireless Dominica to reconnect the 800 lines.

The Battle Is On

Marpin made it clear in the media that legal action would commence soon. This gave a considerable advantage to Marpin, of course, because we were taking the initiative.

Since the law favoured Cable and Wireless, Marpin's task was to find out if there was a higher law being violated. Since there was no computerised legal search system available at that time, a "feet on the ground" search was on, led by a team of Marpin's lawyers: Jenner Armour, Geoffrey Harris, and Anthony Astaphan.

The constitution of the Commonwealth of Dominica at section 10(1) states the following:

> Except with his own consent, a person shall not be hindered in the enjoyment of his freedom of expression, including freedom to

hold opinions without interference, freedom to receive ideas and information without interference, freedom to communicate ideas and information without interference (whether the communication be to the public generally or to any person or class of persons) and freedom from interference with his correspondence." Section 10(2) states, "Nothing contained in or done under the authority of any law shall be held to be inconsistent with or in contravention of this section to the extent that the law in question makes provision:

(a) that is reasonably required in the interests of defence, public safety, public order, public morality or public health;

(b) that is reasonably required for the purpose of protecting the reputations, rights and freedoms of other persons or the private lives of persons concerned in legal proceedings, preventing the disclosure of information received in confidence, maintaining the authority and independence of the courts or regulating the technical administration or the technical operation of telephony, telegraphy, posts, wireless broadcasting or television; or,

(c) That imposes restrictions upon public officers that are reasonably required for the proper performance of their functions.

And except so far as that provision or as the case may be, the thing done under the authority thereof is shown not to be reasonably justifiable in a democratic society.

Since in the opinion of the legal team none of the exceptions applied, Cable and Wireless and the government were acting illegally by issuing the Telecoms Act 1995, thereby interfering with the rights of Marpin to communicate using its own satellite facilities as well as cable and wireless telephone lines for return.

Documents were filed and served on Cable and Wireless and on the government, thus initiating the battle. Knowing that the battle would take quite a while, Marpin accelerated the upgrading of its infrastructure aiming to have full telephone and Internet service available prior to the determination of the case.

However, Marpin was still unable to offer Internet services until it began using its own return lines in early 1999, especially since Cable and Wireless had brought in a telecoms expert to say that such service was not yet economically feasible and was only being used on a trial basis throughout the world. This claim was made despite the fact that research showed that in the city of Rochester, New York, where I was living at the time, telephone and Internet services were being provided to my own home by a cable television company.

Cable and Wireless had a very large team of lawyers headed by Sir Henry B. Ford from Barbados with consultants from the United Kingdom and a local attorney named Alick Lawrence.

The case went through a tug of war with very persuasive arguments made by both the prosecuting attorney and the defence attorney. Listening to the presentations of both sides from the standpoint of a layman, I felt myself being swayed with each skilful argument, even though I was clearly on the side of my own company, Marpin. The presentations were so intense that I could never tell at any one time which side was winning or losing, or whether they were even making valid points.

The Marpin team, headed by Jenner Armour and aided by Anthony Astaphan and Geoffrey Harris, remained confident throughout the trial that Marpin would prevail. Wanting to be involved with the legal team representing my company, I put in long hours seven days a week throughout the trial, assisting with typing, copying, stapling, and making of folders until the bundles were being measured in feet, being in excess of three, so large and extensive was the work.

Food was brought in and breaks were minimal. I remember falling asleep in a chair and then waking up hours later, only to find that the team was still at it, discussing possible arguments and counter arguments to ensure no stone was left unturned.

This carried on for about six gruelling weeks until finally all arguments and exhibits had been made and submitted, and judgement was reserved by the presiding High Court judge, Justice Dunbar Cenac, for a later date, during which time rumours and counter rumours of a verdict were the order of the day.

During this time, while I was away at a conference in Trinidad, a high-ranking Cable and Wireless executive approached me and told me that with all that Cable and Wireless stood to lose in trial, he wondered whether I thought they would leave the decision to chance? That caused me to tremble inwardly for a while, though I didn't show it, especially not in front of this executive, but despite my worries, ultimately I believed that justice would prevail in our favour.

Marpin's Internet service in the meantime was more consistent, faster and cheaper than the competition, but was not yet available throughout the island as upgrade work was still ongoing. This work was being

hampered, however, by our limited funds due to the legal costs and miscellaneous expenses incurred by the lengthy trial.

While all this was going on, Marpin was also testing its voice services to be able to capitalise on local terminations, an aspect of the industry that was also being monopolised.

My company had to sell shares to be able to install the voice infrastructure, which cost us some $6,000,000. This installation had to be done so that after a decision was made (in our favour, we hoped), we could show that Marpin was able to provide the services that we had fought so hard to obtain. This investment was also undertaken because our legal team were so confident of victory in the case.

On April 29, 1999, exactly four years after the monopoly agreement was signed by Dame Eugenia Charles, the decision was given. To this day, I can remember my confusion as Judge Dunbar Cenac read the ruling; when he had finished, I turned to the legal team leader and asked what, exactly, did the ruling mean. He answered, "Ron, you won!"

THE INDEPENDENT

Wednesday May 05, 1999   Vol. III No. XXXXVIII   **Commonwealth of Dominica**   EC$1.50

VICTORY...

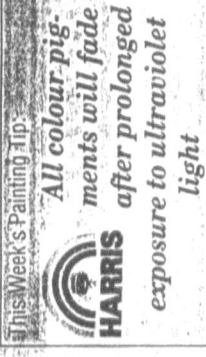

Marpin Telecoms wins court battle against Cable and Wireless

Telecoms and Broadcasting Ltd. over Telecommunications grant Cable and Wireless.

The court battle emerged, after Cable & Wireless disconnected the 1-800 numbers provided to MARPIN Telecoms which enabled the company to provide dial-up access and internet services for its customers in Dominica.

Cable & Wireless explained, at a press conference, that it had take this action, after receiving advice that MARPIN has been acting beyond the scope of its license and therefore committing an offence by providing Internet via a satellite earth station.

MARPIN Telecom in response believed that its 1-800 numbers were wrongfully disconnected and had given Cable & Wireless an ultimatum, either to reconnect its 1-800 numbers or face legal action.

Cable and Wireless failed to reconnect MARPIN's 1-800 numbers and MARPIN Telecoms took the matter to Court.

The Court therefore ordered that the application

for the restoration MARPIN 1-800 numbers be granted. The court has also ruled that the exclusivity conferred by license dated 29th April, 1985, granted to Cable & Wireless pursuant to Section 7(1) of the Telecommunications Act and permitted by Section 9 of the act has been declared to be in contravention of Section 10 (1) of Dominica's Constitution and therefore accordingly is invalid.

The court also ruled that Section 7 (1) of the Telecommunications Act to the extent that the Ministry is prohibited from issuing a licence to any person other than Cable & Wireless under the agreement is in contravention of Section 10 (1) of the Constitution as violation of freedom of expression. The court has also ordered government to pay the MARPIN Telecom's legal cost.

Anthony Astaphan, one of Marpin's legal counsel team informed The Independent that this was a significant and historic case for MARPIN Telecoms. Mr. Astaphan believes that it now sets everything in place for the liberalization of the Tele-

communications market.

Mr. Astaphan said, "we believe the Judge was spot on in this judgement and we are very happy, and "MARPIN Telecoms can go ahead with their plans right now."

And how does Ronald Abraham General Manager of Marpin feel about his company victory? When contacted by *The Independent* "I am very happy and thankful to God, this is the day we have won back for the Commonwealth of Dominica the right to control its telecommunications resource. We've also by extension, since we all have similar constitutions in the region, won a battle for the OECS."

Mr. Abraham also informed *The Independent* that the company is preparing to offer telephone services in June and they have already installed all their equipment. He said that customers are expected to receive equal or better quality service since the services to be provided by MARPIN Telecoms will be digital and that, there definitely will be a reduction in rates. Within the next year international calls are expected to drop by more than half in Dominica, Mr. Abraham also informed *The Independent.*

Efforts by the *Independent* to reach Cable & Wireless General Manager, Mr. Carl Roberts and their attorneys were futile.

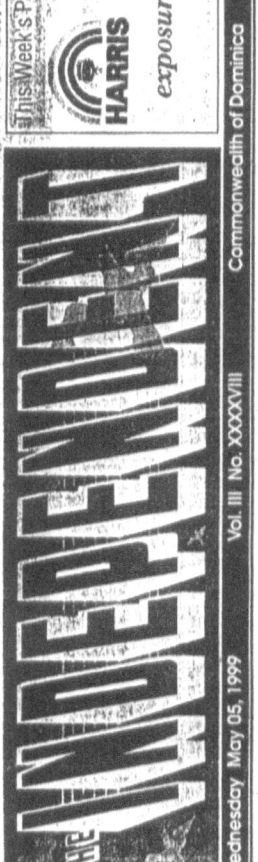

Managing Director of MARPIN Telecoms, Mr. Ronald Abraham

A decision with far reaching consequences could be used to describe the High Court Judge Justice Dunbar Cenac's ruling in favour of MARPIN

There is Hope

CHAPTER 5 – THE FIGHT CONTINUES

A new day had been ushered in by our momentous victory. The monopoly was ruled illegal and therefore, Marpin was allowed to provide full telecoms services. A licence was granted to Marpin to provide full telecom services for 50 years of every kind similar to that provided by Cable and Wireless.

There was great rejoicing throughout the Caribbean that the seemingly impossible had been accomplished. It humbled me greatly that my name, Ron Abraham, had become legendary.

On December 15, 1999, Marpin officially launched local and international telephone services at a rate cheaper than the rate offered by our competitor, and our good rate came with consistently good quality, as well. Furthermore, local Marpin to Marpin calls were free.

An extract from the media story written by *The Independent* captures the positive feelings of the local residents concerning Marpin telephone services.

MARPIN PHONES RINGING

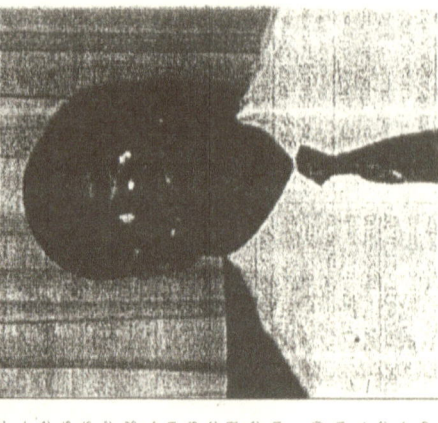

As it launched its new telephone service, Marpin Telecoms and Broadcasting complains that Cable and Wireless - Dominica, which until recently had a monopoly on telephones, is not providing it with a service that would connect the two networks of subscribers. Marpin contends that the delays likely to place Dominicans in an embarrassing situation since they will be unable to call customers using a different network and that Cable and Wireless is hindering Communication between the people. Mr. Ronald Abraham, executive director of Marpin stated that no company has the right to obstruct the communication of the people and the people of Dominica are being obstructed for communicating with each other by not receiving inter-connection. "What I'm seeing from the professional standpoint is a lot of delay tactics, but even that, too, will come to an end," Marpin's Promotion and Marketing Manager, Anestine Theophile-Lafond said.

Mr. Abraham stated that Marpin started working towards achieving telephone services for customers in April 1998. He said that it is the constitutional right of the people to access telephone services through their own platform. With this newly established technology system, customers will be able to access telephone, television and Internet services through one line. This will be achieved using a remote indoor switching unit.

Employees of Marpin are already being connected and connections will commence to customers by December, this year, however these customers will receive partial service. Complete services are expected by June 2000. Mr. Lafond stated that first connections will be for the 'vital areas' of our country such as the hospital, police and fire departments, social security and others. Mr Abraham noted that connection will begin on November 1st but the official launching of the new service is scheduled for December 15. He

declared that for Marpin customers connection of the telephone service is EC$20.00 and non-customers can get connected for EC$150.00 for all three services or any one of the three requested by the customer. A flat rate of EC$30.00 is allotted for unlimited local calls to other Marpin customers and international call rates, according to Mr. Abraham, are comparable to those of the competition. The rate for calling Cable and Wireless customers have not yet been finalized, but he assured that it will not exceed 28 cents. Mr. Abraham exclaimed, "Marpin is not afraid of competition, nor is Marpin afraid of competitive prices Marpin is a dynamic company. We look at the market and adapt. We will deal with the market place as the market place evolves". He urged the government however, to stop predatory pricing in Dominica.

Mr. Abraham declared that Marpin has already invested $40 million in purchasing cellular equipment, about $9million on the telephones and intends to invest a total of $60 million. On connection to the service, customers will receive the remote indoor switching unit, a back-up rechargeable battery and an Internet service device. The battery, according to Mr. Abraham, delivers 4 hours of on time, but only 2 hours of talk time. Customers can, however, request the packages with either the 8 hour or 16 hour battery. He revealed that the prefix for dialing telephone numbers is 500 - 504 in comparison to the Cable and Wireless's 44' prefix.

Mr. Abraham noted that Marpin's next step is to establish digital television, where the customer can decide what programmes or part of programmes they want and pay accordingly. He declared that the Marpin staff has increased from 55 to 100 in the past two years and may be considering increasing by about 30 more. He mentioned that the final phase of the new Marpin building is expected to commence within the next two to three weeks.

Mr. Ronald Abraham,
C.E.O Marpin Telecoms.

"No case to answer"

Dear Sir,

On or around the 19th December 1997, I wo awakened from my bed and told that polic wished me to assist them with the investigation o a report concerning drugs and ammunition I wo taken from my home and brought to another re idence newly built by me where two Puert Rican visitors were renting. My farm worker wo

This victory had a ripple effect throughout the region. Governments were quickly calling for negotiations with Cable and Wireless to end monopolies in light of the Marpin decision.

New telecoms bodies were formed, and in the OECS (Organisation of Eastern Caribbean States), a body of ministers began negotiating with Cable and Wireless to liberalise the market, especially since the constitutions were similar, and as a result, the Eastern Caribbean Telecoms Authority (referred to as ECTEL) came into existence.

Cable and Wireless lodged an appeal with the West Indies Court of Appeal, which to our dismay (but not surprise) was accepted, and the battle continued as before: We launched our defence by searching for fresh supporting cases and preparing bundles of information, but at least this time our efforts did not require the same amount of sleepless nights.

In the meantime, Marpin applied for interconnection with Cable and Wireless vital for competition and was denied on the grounds that nothing could be done until the appeal was heard.

Marpin continued to make inroads in the provisioning of high speed cable modems; the never before available large bandwidth created a spin off industry as people began downloading movies and music videos over the Internet. Other new industries came available as well, such as Internet gaming, offshore banking, and calling centres. At last, Dominica was buzzing with life.

News of recently signed agreements with Cable and Wireless in Jamaica, Trinidad, Grenada, Barbados, St Kitts and Nevis opened up the market even more, and consortiums started to plan fibre links throughout the region.

As these events were unfolding, we at Marpin waited for the judgement to be handed down from the Caribbean Court of Appeals; when it was, we were ecstatic. The decision of the Court of first instance was upheld.

This spurred us to seek interconnection from Cable and Wireless, but due to a rift that was becoming evident between Marpin and the Minister of Communication and Works (though this was publicly denied by the Minister), no help or assistance of any kind was given to Marpin by the government in the battle for interconnection. This technically went against the new Telecoms Act 2000, which states in section 12 (1) subsection k that commission should do so; furthermore, section 46 states, "…a telecommunications provider who operates a public Telecommunications network shall not refuse, obstruct, or in any way impede another telecommunications provider from making an interconnection with his telecommunications network."

Since some members of Marpin's board of directors supported the government, the board was divided, but as CEO, I was able to obtain a majority vote and muster general agreement for a public demonstration asking for interconnection capabilities.

The battle for interconnection went as far as a meeting with the OECS ministers of Government in Grenada on September 21, 2001, at Rex Grenadian Hotel in Grenada.

Given below is a detailed account of this meeting based on my notes taken while in attendance.

Report on the meeting called by the Organisation of Eastern Caribbean States (OECS) Secretariat on the current status of Interconnection negotiations between Marpin and Cable and Wireless, held at the Rex Grenadian Hotel on September 21, 2001 in Grenada.

The following attended the meeting:

Dr. Keith Mitchell – Chairman OECS/ Prime Minister of Grenada **(KM)**

Ms Jennifer Astaphan – OECS Secretariat **(JA)**

Mr Phillip Lacorbiniere – OECS Secretariat **(PL)**

Mr Geoff Batstone – Cable and Wireless **(GB)**

Mr Trevor Clarke – Cable and Wireless (**TC**)

Mr Errard Miller – Cable and Wireless (**EM**)

Ms Lisa Agard – Cable and Wireless (**LA**)

Mr Reginald Austrie, Minister – Communications & Works Dominica (**RA**)

Mr Eluid Williams, Permanent Secretary – Communication & Works Dominica (**EW**)

Mr Ronald B. M. Abraham – Marpin Telecoms (**RBMA**)

Mr Joffrey Harris, Senior Counsel – Marpin Telecoms (**JH**)

Mr Duncan Stowe – Marpin Telecoms (**DS**)

The meeting began at about 1:10 pm with the Chairman, Dr. Keith Mitchell (**KM**), calling the meeting to order. The Chairman explained that he and the OECS were concerned about global events in the industry, particularly in the US, and the impact on the OECS of those events. He noted that Dominica was under severe economic pressure and with the reduction in the growth and export of bananas, it needed an alternative means of economic survival. He said that the telecommunications sector would have to play a crucial role in assisting the development of Dominica and the OECS. With that in mind, **KM** said that he felt it was important to invite both Marpin and Cable and Wireless to the present meeting so that both parties could enlighten the OECS Secretariat on the status of the interconnection negotiations between the two companies.

Cable and Wireless was called to speak first.

GB told the meeting that the two companies had met on several occasions and had made considerable progress toward attaining the goal of interconnection. He referred to the fact that three (3) documents had already been approved and signed by both parties. He stated further that the other four (4) documents had already been sent to Marpin by

Cable and Wireless and they were awaiting approval by Marpin. Much of the presentation by **GB** came from a text prepared by Cable and Wireless.

On behalf of Marpin, I **(RBMA)** thanked the OECS Secretariat for facilitating the meeting. I noted that notwithstanding the fact that the parties were at the negotiating table, Marpin wished to place on record the following objections/issues:

> **Objection:** Marpin objected to the fact that Cable and Wireless was only inclined to grant Marpin domestic interconnection rather than full interconnection. He stated that the refusal of Cable and Wireless to grant full interconnection was contrary to Section 46 (4) of the Dominica Telecommunication Act 2000.

LA of Cable and Wireless insisted that C & W would be honouring clause 8 of the Memorandum of Understanding (MOU) between C & W and the contracting OECS governments and was therefore only willing to grant Marpin domestic interconnection under the interim Interconnection Agreement.

> **Objection:** Although Marpin had signed an Indemnity Agreement to facilitate the early ordering of interconnection equipment from Cable and Wireless, Marpin was very unhappy with a condition in the agreement that stipulated that Cable and Wireless would keep all the interconnection equipment in boxes until all documents were signed by both parties. **RBMA** said that such a condition in the Indemnity Agreement could lead to unreasonable delays in the interconnection process.

DS also stated that Cable and Wireless' reservations about selling the equipment if interconnection was not realised within twelve months of the purchase date cast considerable doubt on Cable and Wireless' willingness to conclude the interconnection process with Marpin. He said that the latter term did not show a sign of good faith by Cable and Wireless.

LA remarked that Cable and Wireless was only following basic business practices, and this did not mean that Cable and Wireless was bargaining in bad faith.

> **Objection:** Marpin objected to the Introduction of Access Deficit Contribution required by Cable and Wireless in the draft Tariff Schedule.

PL of the OECS Secretariat stated that Cable and Wireless could not charge an Access Deficit Contribution unless it was approved by ECTEL and referred to clause 20 (ii) of the MOU.

Minister Reginald Austrie **(RA)** stated that he was disappointed with the fact that interconnection negotiations were taking place between the two companies, and yet he had no access to information on the state of negotiations due to a non-disclosure agreement. He further stated that both his ministry and the government were currently under public pressure due to the lack of of information on the Interconnection process.

Minister **(RA)** told the meeting that the Dominican public had been clamouring for interconnection between Marpin and C & W for over three years. He said he was not comfortably leaving Grenada without having set a firm date by which interconnection would occur.

EM of C & W stated that Cable and Wireless was willing to work toward October 30, 2001, as the date by which interconnection would take place. He noted, however, that Cable and Wireless would not commit itself to October 30 being the final date.

On the question of the public information clause in the Non-disclosure Agreement, **JH** of Marpin said that Marpin was willing at any time to withdraw the said clause from the agreement if Cable and Wireless was also willing to do so.

At the end of the meeting, **GB** of C & W sought and received Marpin's consent for copies of a document from C & W, which Marpin had never seen before, to be presented to the Minister **(RA)**. However, when the document was presented to **RA**, Marpin made it clear that the document had been prepared solely by C & W, and that Marpin had not been involved in any way. Thus, should any public statement be made by the Minister or the OECS Secretariat from the said document, it should be made clear that it was a C & W-released document.

> **Objection:** Having read the document, Marpin noted that although C & W had agreed during the meeting to work toward October 30 as the date for interconnection, the document stated November 2001 as the target date.

Both Marpin, and Cable and Wireless agreed that sometime in the future, they would make a joint public announcement on the status of interconnection.

About this same time, the high court appointed Jenner Armour, Anthony Astaphan and Geoffrey Harris as Senior Counsels for their collective part in winning such a historic case.

Due to the large increase in routine legal work my company faced, we (Marpin) obtained the services of a junior counsellor named Duncan Stowe, who was the main force behind the organisation of the demonstration for interconnection when hundreds of people took to the streets of Roseau, the capitol city, demanding interconnection *now*.

The heads of government empowered ECTEL to ask for tenders for telecoms services with a grandfather clause allowing Marpin and Cable and Wireless to receive their licences first before all others, but although we (Marpin) applied for mobile services, when our licence was issued, the lucrative field of mobile services was denied us.

The article transcribed below gives an insight by the *SUN* newspaper of Dominica on the rejection Marpin received when he applied for a mobile licence application.

MARPIN'S REJECTION – IS IT PERSONAL?

With Marpin Telecoms and Broadcasting Company Limited (Marpin) having failed in yet another bid to secure a licence from the Eastern Caribbean Telecommunications Authority (ECTEL) to operate a mobile phone service, officials of the local company have again taken the attack to the Minster of Communications and Works, Reginald Austrie.

"It is personal," Marpin spokesman Duncan Stowe told the *SUN*. "I am confident that if any other government comes into office, we will get our licence."

Stowe was convinced that the minister "has never forgiven us for comments (that questioned the integrity of the minister) we made about him" during several months of "agitation" by Marpin for a licence to operate a fixed line telephone service; a sentiment also expressed by Ron Abraham, Marpin's chief executive officer.

"During our agitation we shared with Dominicans certain information. The minister took it personally and is making Marpin pay for it," Abraham contended.

Minister Austrie has flatly denied Marpin's claims, repeating an often-stated position that applications for telecommunications licences go to ECTEL for processing after going through the National Telecommunications Regulatory Commission (NTRC), and that the minister is only advised by the NTRC after a decision has been taken on whether or not to grant a licence.

In June 2000, five Eastern Caribbean states – Dominica, Grenada, St. Kitts/Nevis, St. Lucia and St. Vincent and the Grenadines – passed harmonized telecommunications laws that paved the way for the establishment of ECTEL.

Part three of the Act states that after a person has applied for a licence, the application will be reviewed by ECTEL, which will make the appropriate recommendation to the minister through the local telecommunications commission.

"The Minister in deciding whether or not to grant a licence will take into account ECTEL's recommendation, the purpose of the Treaty and the public's interest," it states.

"Marpin can only claim unfairness if ECTEL recommends (that they be granted a licence) and the minister does not grant it. To date, for all the times Marpin have applied and reapplied, they have always failed the criteria (and) if ECTEL says no, I can only act accordingly," Austrie told the *SUN*.

In order to be granted a licence, the applicant must satisfy three key requirements – legal, technical and financial, ECTEL has stated.

While Marpin has satisfied the first two, the local company has been unable to convince the evaluators that it has or can raise the necessary funding to offer a mobile phone service.

And in what appears to be a twist of irony, the company that fought for the liberalization of the telecommunications sector in Dominica, and by extension the Eastern Caribbean, appears to be too broke to take full advantage of liberalization.

"Things are very tough (with us financially)," Abraham, also a director, admitted to the *SUN*, although he refused to give details of the company's financial position.

But the *SUN* has learned that the company has outstanding debts of over EC$10 million, including over EC$4 million to Barclays Bank and over EC$5 million to the National Commercial Bank.

The situation is so bad that Barclays Bank has cut Marpin's overdraft facility by 50 percent and financial institutions have refused to grant the company any loans, admitted Abraham who, as CEO, receives a monthly salary of EC$14,000 in addition to allowances. His wife who is manager of broadcasting, is paid about EC$6000, according to Duncan Stowe, although other sources have told the *SUN* that it was closer to EC$10,000.

Stowe, an attorney, has defended Abraham's salary insisting that the CEO was extremely valuable to the company.

"There isn't another person in Dominica that can run Marpin the way that Ron (Abraham) does. He really knows the terrain. To replace Ron you will need about three people, or four," Stowe asserted.

In any event, he said, a "particular director" had received over a million dollars in brokerage fees over the years "because he is a guarantor."

In an attempt to meet the financial requirement that ECTEL demanded after turning down a previous application, Marpin entered a joint venture arrangement with a US-based company, Pan Caribbean Holdings, and formed a new company solely for the purpose of offering a mobile phone service.

Pan Caribbean Holdings would own 65 percent of the new company called Marpin Wireless, and Marpin Telecoms and Broadcasting would contribute its licence for a 35 percent share, something Austrie said the local company could not do because the licence remained the property of the government.

ECTEL rejected the submission stressing that it was Marpin Telecoms and Broadcasting and not Marpin Wireless that had applied for a licence; that Marpin Wireless had not met the January 30, 2002 deadline for submitting an application for a licence; and that Marpin Telecoms and Broadcasting had again failed to provide the necessary information to satisfy the regulators that it was financially viable.

"Due to the fact that Marpin Wireless is a separate and distinct legal entity MTBC (Marpin Telecoms and Broadcasting Co. Ltd) cannot lawfully rely on the purported strength of Marpin Wireless in order to satisfy the financial requirements of the evaluation process," then managing director of ECTEL, Donnie DeFreitas wrote to the chairman of the NTRC, Dr. Nicholas Liverpool.

In the letter dated October 4, 2002, a copy of which was obtained by the *SUN*, DeFreitas emphasized that it was Marpin that was required to satisfy the evaluators of its sound financial standing and that it had failed to do so.

"It cannot attempt to call to its aid the unsubstantiated source of funds from an unproven entity that is at best, separate and distinct," he wrote. "Accordingly, ECTEL recommends that MTBC be not granted a Public Mobile Telecommunications Licence."

In a letter dated October 14, 2002 the minister of communications relayed the ECTEL decision to Abraham.

The Marpin CEO told the *SUN* he would reply to the minister soon but he refused to divulge what he would state in his letter.

PRIVY COUNCIL

Cable and Wireless appealed to the Privy Council of London, the final court of appeal for the Caribbean, and although it was felt there was no ground for further appeal, they applied by special licence for an appeal hearing and leave to appeal was granted.

While this was going on, Senior Counsel Jenner B.M Armour fell ill and died; thus, his major and vital role in that case ended a great and sad loss for the Marpin cause.

Since leave to appeal had already been granted, Marpin had no choice but to pursue the matter.

In a bid to make up for the great loss of Senior Counsel Jenner Armour, Sir Carl Hudson Phillips from Trinidad, at the recommendation of Senior Counsel Anthony Astaphan, was added to the team.

As the CEO of Marpin, my confidence was greatly affected at losing the tremendous role and depth of legal experience that Senior Counsel Jenner Armour had brought to the table.

Meanwhile, our legal fees had sharply increased because a legal consultant with Privy Council experience needed to be added to our team, and then another senior lawyer from England had to be added as well.

The decision was taken by Marpin and its legal team to allow Senior Counsel to lead the Marpin case. In retrospect, that was a serious error of judgement, although one is unable to say for sure what impact it would have had on the case.

I believed the case for our company would have been better served had we secured the services of a QC (Queen's Counsel) from England, someone who was well versed in presenting cases at that level.

In May 2000, the team travelled to London with Senior Counsels Astaphan, Harris and Sir Carl Hudson Phillips. As the CEO, I funded their expenditure of travelling in first class, while I travelled in tourist class. The team stayed at the Sir Winston Churchill Hotel in Mayfair where work continued along with the British Counsel in preparing the case.

When it was time to appear before the Privy Council, the Cable and Wireless case was presented by Lord Lester. When it was Marpin's turn, Senior Counsel Astaphan stood up to make his presentation. While looking on and observing how Astaphan conducted himself, I felt that he was overwhelmed by the occasion at first, but in time, he regained his composure.

Once again, the to and fro volley of arguments and precedents continued unabated, but I felt very comforted to be supported by the presence of

Ivor Nassief, a young businessman from Dominica who travelled at his own expense to be present for the case at Privy Council.

After three days, the case was brought to a close. The Marpin legal team broke up and went their separate ways leaving, and I remained in London. The judgement was handed down a few weeks later by Lord Cooke of Thorndon, a copy of which is reproduced in full below:

CABLE AND WIRELESS (DOMINICA) LIMITED
v.
MARPIN TELECOMS AND BROADCASTING COMPANY LIMITED
(DOMINICA) [2000] UKPC 42 (30TH OCTOBER, 2000)

Privy Council Appeal No. 15 of 2000

Cable and Wireless (Dominica) Limited, *Appellant*

v.

Marpin Telecoms and Broadcasting Company Limited,
Respondent

FROM

**THE COURT OF APPEAL OF THE
COMMONWEALTH OF DOMINICA**

JUDGMENT OF THE LORDS OF THE JUDICIAL
COMMITTEE OF THE PRIVY COUNCIL,
Delivered the 30th October 2000

Present at the hearing:

Lord Nicholls of Birkenhead

Lord Steyn

Lord Cooke of Thorndon

Lord Clyde

Lord Hobhouse of Woodborough

*[Delivered by **Lord Cooke of Thorndon**]*

1. The issue in this appeal is whether an exclusive licence to provide national and international telecommunication services in, to and from the Commonwealth of Dominica infringes that country's constitutional guarantee of freedom of communication.

2. The appellant, Cable and Wireless Dominica Limited (CWD), holds an exclusive licence to provide such services. The licence does not extend to broadcasting. It is for a term of 25 years and was granted by the Minister under the Telecommunications Act 1995, enacted on 26th April 1995. Although the licence was not issued until shortly after the Act came into force, namely on 29th April 1995, both the licence and the Act itself implemented heads of agreement between the Government and Cable and Wireless (West Indies) Limited (CWWI), dated 23rd March 1995.

3. CWWI had provided the international telecommunication service for Dominica since about 1929 and the internal service since about 1967. Since September 1985 it had held an exclusive 20 year licence covering both national and international services. The Government had no shares in CWWI. A main purpose of the heads of agreement in 1995 was to enable the Government to become the holder of 20 per cent of the shares in the new company, CWD, to be formed to take over the services. The Government was also entitled to royalties, and the capital for its shares was to be found by a cash advance to be repaid out of future royalties.

4. The respondent Marpin Telecoms and Broadcasting

Limited (Marpin), formerly Marpin TV Company Limited, began cable television operations in Dominica in 1983. Currently it holds a licence issued by the Minister on 1st March 1996 under the Act of 1995 and authorising it to install, maintain and operate a television station and related telecommunication services. These operations evidently do not compete with those of CWD. Marpin wishes to compete with CWD, however, in the provision of public telecommunication services, particularly at the present stage mobile telephone services and e-mail and internet services offering international communications.

5. In early 1996 CWD was advised by the Minister that Marpin's licence entitled Marpin to provide internet services utilising CWD's network. In January 1997 CWD entered into an internet service provider (ISP) agreement with Marpin whereby Marpin acquired access to the internet via leased lines and terminating equipment supplied by CWD; toll free 1-800 numbers were allotted by CWD, enabling the customers to have internet access. But in March 1998 Marpin gave notice to CWD that it would be terminating the ISP agreement. Instead of using the leased circuits, Marpin used VSAT (very small aperture terminal earth station). This enabled traffic to be sent to orbiting satellites and relayed to receiving earth stations in other countries for onward transmission, and *vice versa*. Thus Marpin was able to bypass a major part of the CWD network, ceasing in that respect to be a reseller of CWD's network services. In retaliation CWD withdrew the 1-800 numbers.

6. By notice of motion dated 20th October 1998 Marpin sought declaratory and other relief under section 16 of the Constitution. CWD was named as the first respondent, the Attorney-General of Dominica as the second respondent. The proceedings challenge the validity of the Act and the licence, insofar as the Act authorises and the licence grants the exclusive licence issued to CWD; yet the Attorney-General

has taken no active part in the proceedings and was not represented before their Lordships' Board. A suggested reason is put forward in the case for CWD, wherein it is said that the litigation has substantial implications for other jurisdictions in the Caribbean and beyond. "In many cases", it is alleged, "it may be of immediate financial benefit to governments, as well as to potential competitor companies, for such countries to be relieved of their existing licence obligation to the incumbent company (because government thereby acquires an unlooked-for power to grant competitive licences)." To this it should be added that another factor may well be the increasing international recognition of the desirability of fair competition in the telecommunications field, sometimes after initial phases of monopoly.

7. The present policy or motivation of the Government of Dominica is not a matter which on this appeal their Lordships either need or could investigate. They are concerned solely with the effect of the Act of 1995 and the exclusive licence in the light of the relevant provisions of the Constitution. Nor does the appeal turn on any narrow questions of the detailed wording of the Act or the licence. Accordingly it is unnecessary for their Lordships to reproduce the wording at any length, as was helpfully done in the judgments below. Some particular terms of the licence require quotation later, but at this point it is sufficient to say that the general effect of the Act and the licence, assuming the validity of both, is clearly to confer on CWD an extensive monopoly in telephonic and other telecommunication services in Dominica for 25 years.

8. Marpin's constitutional motion was heard by Cenac J. in the High Court of Justice over eight days in March 1999. In a judgment delivered on 29th April 1999 the judge granted the application. He made declarations that, in short, the exclusivity conferred by CWD's licence of 29th April 1995 was in contravention of section 10(1) of the Constitution, and accordingly invalid; and likewise that section 7(1) of the

Act, to the extent that the Minister is prohibited from issuing a licence to any person other than CWD, is in contravention of section 10(1) of the Constitution, and accordingly invalid. He also ordered that Marpin's costs be paid by the Attorney-General. On the constitutional issue CWD appealed from that judgment. Marpin cross-appealed on the costs question. The case came before the Eastern Caribbean Court of Appeal (Singh and Redhead JJ.A. and Matthew Ag. J.A.) on three days in September 1999. By a judgment delivered by Redhead J.A. on 8th November 1999, the Court of Appeal dismissed the appeal, agreeing substantially with Cenac J. on the constitutional question, but allowed the cross-appeal, ordering that CWD pay Marpin's costs in both courts. CWD now appeals by special leave granted by the Judicial Committee.

THE CONSTITUTIONAL PROVISIONS

9. Section 1 of the Constitution provides:

"Whereas every person in Dominica is entitled to the fundamental rights and freedoms, that is to say, the right, whatever his race, place of origins, political opinions, colour, creed or sex, but subject to respect for the rights and freedoms of others and for the public interest, to each and all of the following, namely:

  a) life, liberty, security of the person and the protection of the law;

  b) freedom of conscience, of expression and of assembly and association; and

  c) protection for the privacy of his home and other property and from deprivation of property without compensation, the provisions of this Chapter shall have effect for the purpose of affording protection to those rights and freedoms subject to such limitations of that protection as are contained in those provisions, being limitations designed to ensure that the enjoyment of the said rights and freedoms by any person does not prejudice the rights and freedoms of others or the public interest."

10. The concept of freedom of expression is enlarged as well as enshrined in section 10 of the Constitution:

1) Except with his own consent, a person shall not be hindered in the enjoyment of his freedom of expression, including freedom to hold opinions without interference, freedom to receive ideas and information without interference, freedom to communicate ideas and information without interference (whether the communication be to the public generally or to any person or class of persons) and freedom from interference with his correspondence.

(2) Nothing contained in or done under the authority of any law shall be held to be inconsistent with or in contravention of this section to the extent that the law in question makes provision:

    a)   that is reasonably required in the interests of defence, public safety, public order, public morality or public health;

    b)   that is reasonably required for the purpose of protecting the reputations, rights and freedoms of other persons or the private lives of persons concerned in legal proceedings, preventing the disclosure of information received in confidence, maintaining the authority and independence of the courts or regulating the technical administration or the technical operation of telephony, telegraphy, posts, wireless broadcasting or television; or

    c)   that imposes restrictions upon public officers that are reasonably required for the proper performance of their functions, and except so far as that provision or, as the case may be, the thing done under the authority thereof is shown not to be reasonably justifiable in a democratic society."

HINDRANCE

11. The Constitution thus treats freedom of expression as including freedom to receive and communicate ideas and information without interference. Except with his own consent, a person is not to be hindered in the enjoyment of this freedom. The first question in the present case is accordingly, under section 10(1), whether Marpin's freedom to communicate ideas and information through telecommunications is hindered by CWD's monopoly. To that question their Lordships think, in company with the courts of Dominica, that the answer can only be in the affirmative. The extent of the hindrance and its reasonableness or otherwise are for consideration under section 10(2). The degree of efficiency of CWD's services and the level of charges imposed by CWD with the approval of the Minister (required by the licence) are similarly relevant under that subsection. But some significant hindrance to a would-be competitor's freedom is normally inherent in any requirement that he provide to his customers certain services only if permitted and on terms laid down by a monopolist. There is no ground for putting this case into any exceptional category.

12. There is a dearth of case law directly in point, but such authority as there is supports the foregoing conclusion. In the leading European case of *Autronic AG v. Switzerland* (1990) 12 E.H.R.R. 485 the applicant was a private commercial Swiss company specialising in home electronics. It applied for permission to receive, by means of a private dish aerial, uncoded television programmes intended for the general public from a Soviet telecommunications satellite, the

company's object being to give demonstrations of the technical capabilities of the equipment in order to promote sales. The satellite provided a fixed point-to-point radio communication service. It also transmitted telephone conversations, telexes or telegrams and data. In the absence of consent from the broadcasting state, the Swiss authority refused the application. The applicant successfully complained of a violation of Article 10 of the European Convention for the Protection of Human Rights and Fundamental Freedoms (1953) (Cmd. 8969), corresponding broadly though not exactly to section 10 of the Dominican Constitution. The court held that there had been an interference with the company's right to receive information, saying in the course of paragraph 47 of the decision:

Article 10 applies not only to the content of information but also to the means of transmission or reception since any restriction imposed on the means necessarily interferes with the right to receive and impart information.

13. The court further held that the interference was not "necessary in a democratic society." That part of the decision is relevant at a later stage of the present judgment. The part from which the previous quotation is taken is relevant at this stage and to the reach of section 10 of the Constitution because, although *Autronic* was primarily a broadcasting case, it underlines that freedom of communication is not limited to the information or ideas which a person wishes to convey. The content of the Soviet programmes was immaterial for the technical purposes of the Swiss company. Hence the Case lodged for CWD in the present appeal included a concession that freedom of expression may protect the transmission of information for commercial purposes or profit. Interference with the provision of a telecommunication service, such as that provided by Marpin, can amount to interference with the freedom of expression of those who would wish to use that service.

14. An authority quite closely in point is the *Retrofit* case in Zimbabwe. A company wishing to establish a mobile cellular telephone service successfully challenged the statutory monopoly of the state-owned Posts and Telecommunications Corporation in the provision of public telecommunication services within, into and from Zimbabwe. There are two unanimous decisions of a Supreme Court of five judges, both delivered by Gubbay C.J. Again the constitutional provisions were broadly similar. In the first decision it was held that the monopoly infringed the right of freedom of expression and went further than was reasonably justifiable in a democratic society: *Retrofit (Pvt) Ltd. v. Posts and Telecommunications Corporation* [1996] 4 L.R.C. 489; 1995 (2) Z.L.R. 199 (S). In the second, a rule nisi having been issued calling upon the responsible Minister to show cause why, in relation to mobile telephone services, the monopoly should not be declared unconstitutional and invalid, the court dealt further with the issue of reasonable justification in a democratic society and made a declaration of invalidity: *Retrofit (Pvt) Ltd. v. Minister of Information, Posts and Telecommunications* [1996] 4 L.R.C. 512.

15. The first of those two *Retrofit* decisions includes a survey by Gubbay C.J. of jurisprudence, mainly American, concerning the value of freedom of expression, mainly in broadcasting. The cases cited by the learned Chief Justice will be helpful as to the right approach when issues under section 10(2) of the Constitution of Dominica are being determined. As to the scope of section 10(1), their Lordships would adopt the following proposition in his judgment (1995 (2) Z. L.R. (S) at 216):

These cases, and there are others, underline the principle that restriction upon or interference with the means of communication, whatever form it may take, abridges the guarantee of freedom of expression. A fortiori any monopoly which has the effect, whatever its purpose, of hindering the right to receive and impart ideas and information, violates the protection of this paramount right.

16. Later in the same judgment he says (at 218):

In my view, it is axiomatic that for the corporation to monopolise telecommunications services in Zimbabwe, and then to furnish a public switched telephone network of dubious worth, available to but a small percentage of the populace, manifestly interferes with the constitutional right of every person in the country to receive and impart ideas and information by means of this 'pervasive two-way communications system'.

17. Thus it is apparent from the judgment that the monopoly service in Zimbabwe had shortcomings which may have made that case relatively easy to decide. It is to be noted, however, that the previous more general proposition stated by Gubbay C.J. was not limited to cases of inefficient monopolies. Their Lordships regard the efficiency or otherwise of the monopoly as among the matters falling for consideration under section 10(2) of the Dominican Constitution.

18. Notwithstanding the concession in the appellant's Case already mentioned, the argument for the appellant under section 10(1) was to the effect that Marpin had no constitutional or other right to operate under its own system. In developing this argument Lord Lester of Herne Hill Q.C. relied heavily on the judgment at first instance of Costello J. in the Irish case of *A.G. v. Paperlink* [1984] I.L.R.M. 373. When that decision is examined their Lordships do not consider that it can bear the weight sought to be placed on it. The individual defendants, who through their company operated a courier service in Dublin, were held to have infringed the state statutory postal monopoly. They pleaded a constitutional right of citizens, not express but implied, to communicate freely with one another. The learned judge held that it was not correct, and indeed could be seriously

misleading, to suggest that the defendants enjoyed a right to communicate "freely" (see page 382). He thought that a right to communicate is inherent in the human personality but that a right of *free* communication could not be derived from the Irish Constitution. In Dominica, by contrast, freedom of communication is explicitly guaranteed: *Paperlink* is manifestly distinguishable. It would be inappropriate for their Lordships to embark on a discussion of a further part of Costello J.'s judgment wherein he held that, although the right to earn a livelihood was to be derived from the Constitution, the result was not to require the state to justify the existence of a public monopoly.

19. The basic weakness of the argument for CWD on section 10(1) is that it minimises the importance of the provision that a person shall not be *hindered* in his enjoyment of the rights there specified. As already mentioned, in the view of their Lordships some significant hindrance to freedom of communication is normally and in this instance inevitable if there exists a statutory monopoly to control means of communication as important in the world of today as the telephone. The issue therefore shifts to section 10(2).

LIMITATIONS

20. On this appeal CWD does not rely on either paragraph (a) or paragraph (c) of section 10(2) of the Constitution. The expression "public order" in (a) has connotations of the stability of the state and is not so wide as to encompass the considerations of public interest claimed by CWD to be decisive; while (c) is obviously irrelevant. Nor is it suggested that the exclusive licence of CWD was reasonably required for regulating the *technical* administration or the *technical* operation of telephony; so that part of paragraph (b) is relied on no longer.

21. What is invoked by the argument, described by Lord

Lester as the heart of the appeal, is the reference in paragraph (b) to the rights and freedoms of other persons, read together with the corresponding references in section 1 and the general references in that section to the public interest. It is said that the Constitution calls for a balancing exercise, a complex value judgment upon which the courts should defer to the opinions of the legislature and the executive. It is contended that the latter arms of the state had to decide whether regulated exclusivity or regulated competition was the best system of telecommunications control for Dominica at the present stage of national development. Emphasis is laid on the smallness of the population (about 75,000) and the mountainous terrain. The difficulties of providing a universal telephone service and the importance of committing an international telecommunications enterprise to Dominica are put forward as reasonably justifying a monopoly able to cross-subsidise its services.

22. In support of the argument certain terms and conditions of the exclusive licence were pointed out. Among them clause 3 should be quoted:

> 3. Throughout the term of this Licence and subject to the provisions of clause 19, the Company shall operate, maintain in proper working order, expand and improve such Relevant Telecommunication Services as it operates from time to time with a view to providing an efficient and reliable service over as wide an area of the Territory as may be practicable and in accordance with the needs of the Territory, provided that the Company is satisfied that it has the financial resources with which to meet the required additional expenditure and that such expenditure is commercially justifiable taking into consideration the returns derived overall by the Company from the provision of Telecommunication Services in the Territory.

23. A large body of evidence, consisting of affidavits and cross-examination, was presented by the contesting parties in the High Court. It included the expert evidence for CWD of Professor Hausman, who helpfully traversed the options for telecommunications control available to a country and the material factors. As a whole the evidence was characterised by Cenac J. as "effusive nonetheless instructive." Much of it related to the standard of service provided by CWD to Marpin, the prices quoted or charged to Marpin by CWD, and the present and future level of competence of CWD's and Marpin's operations. But the learned judge dismissed this evidence as irrelevant to the issues in the case. Hence the Court of Appeal and their Lordships have been without the benefit of factual findings by the trial judge on matters seen by the parties as important.

24. In the event both the High Court and the Court of Appeal virtually put aside the main argument for CWD – Cenac J. on the ground that "the rights and freedoms of other persons" cannot apply to the regulation of telephony or telecommunications, as otherwise the words "regulating the technical administration or the technical operation of telephony" in section 10(2)(b) would be redundant; the Court of Appeal on the ground that, in the words of Redhead J.A., "I think that it defies logic and common sense to say that a subsection which limits the rights of a person in one section would give those very rights limited by that subsection to a third person."

25. Their Lordships think that this constitutional issue calls for a rather broader approach. They reject a submission that, in the absence of a relevant cross-appeal, Marpin is debarred from challenging CWD's argument on the merits. They are far from saying that it cannot be effectively answered on the merits. But they hold that CWD is entitled to a consideration of it on the merits.

26. While much of what is said by Gubbay C.J. in his two *Retrofit* judgments will be highly relevant in determining the issue on the merits, there is one observation which might not apply to the situation in Dominica. In the second of two judgments Gubbay C.J. says, [1996] 4 L.R.C. at 516:

A government committed to the grant of affordable telephonic communication for its people in the rural areas must be prepared to bear a portion of the expense required to promote such a commendable endeavour. The remedy lies in subsidising this social need, not in impacting upon a fundamental human right.

27. Their Lordships cannot on the present appeal rule out any possibility of success for an argument that the economic and other circumstances of Dominica may make a monopoly cross-subsidising its services reasonably required for the purpose of protecting the rights and freedoms of the people to communicate freely. The claim that the preservation of a CWD monopoly would be likely to result in significant extensions of telephone services to rural and remote areas not already served was treated with some scepticism in the argument of Mr. Astaphan S.C. for Marpin. The proviso in clause 3 of the CWD licence to the effect that the expenditure must be commercially justifiable in the eyes of the company is hardly reassuring. But in principle the "rights and freedoms of other persons" in section 10(2)(b) of the Constitution is capable of covering the facilities for communication available to the community as a whole. The immediately preceding word "reputations" is not enough, in this constitutional instrument, to warrant a reading down of the scope of "rights and freedoms."

28. On the other hand the argument for the appellant goes too far if it suggests, as it apparently did, that any consideration seen by the legislature or the executive as bearing on the public interest may be advanced to bring a law within section

10(2)(b). The right to freedom of communication would be a fragile thing if it could be overridden by general political or economic policy. So, too, the stress placed by Lord Lester on the need for judicial restraint cannot be allowed to discourage the courts from a firm performance of their proper constitutional role. The true position is stated by the European Court of Human Rights in the *Autronic* judgment, paragraph 61, in words adaptable to the Dominican Constitution:

61. The Court has consistently held that the Contracting States enjoy a certain margin of appreciation in assessing the need for an interference, but this margin goes hand in hand with European supervision, whose extent will vary according to the case. Where, as in the instant case, there has been an interference with the exercise of the rights and freedoms guaranteed in paragraph (1) of Article 10, the supervision must be strict, because of the importance of the rights in question; the importance of these rights has been stressed by the Court many times. The necessity for restricting them must be convincingly established.

29. The Eastern Caribbean Court of Appeal regarded the Government's decision to grant CWD the exclusive licence as motivated by business considerations. They attached importance also to Professor Hausman's evidence as confirming that CWD in seeking exclusivity would naturally have in mind protecting its capital investment. In the opinion of their Lordships the fact that the Government and CWD had a common financial interest in exclusivity does not preclude a claim that it was reasonably required for the purpose of protecting the rights and freedoms of other persons. It does militate against over-cautious judicial deference in scrutinising the claim, for it suggests that protection of such rights and freedoms may not have been the dominant purpose.

30. In the end, however, the question for the court is the objective one whether, in authorising and granting exclusivity,

the Act and the licence make provision that is reasonably required for the purpose of protecting the rights and freedoms of other persons. If that is shown, the onus falling on those who support exclusivity, the burden will shift to Marpin to show in terms of the last limb of section 10(2) that it is not reasonably justifiable in a democratic society. In considering these issues it is to be borne in mind that the telephone plays a key role in the modern community. The only pervasive two-way method of communication at a distance, it is crucial in business, in providing information to citizens, and in the ordinary conduct of daily life. An important question will be whether, on balance, to allow Marpin to compete with CWD will or will not conduce to providing Dominica with telecommunications services giving best effect to the rights of users to freedom of communication.

REMISSION

31. Their Lordships are driven to hold that, whether or not the results reached in the courts below were right, they were reached after an over-circumscribed approach. It would be unsatisfactory for the Board to attempt to resolve the issues without a local evaluation of the evidence based on the correct principles. The case is one calling for an appreciation of local conditions. With the regret that must accompany a prolongation of the proceedings, the case must be remitted to the learned trial judge for reconsideration in the light of the principles set out in the present judgment.

32. Lord Lester suggested that, in that event, there should be a full rehearing. It may be, however, that this can be avoided. It will be a matter for the discretion of the judge, but one course open to him would be to decide on the basis of the evidence already given, supplemented by any updating evidence which he may give leave to adduce.

33. Accordingly, the appeal will be allowed. The decisions of

the Eastern Caribbean Court of Appeal and the High Court of Justice of Dominica will be set aside, and the case will be remitted to the High Court for reconsideration.

34. Marpin must pay the costs of CWD in the Privy Council. The costs orders in the courts below will be set aside. The costs of all proceedings in the High Court and the Court of Appeal, including the costs of the further hearing, are to be as ordered by the trial judge in the light of the outcome of the reconsideration. Their Lordships agree with the Court of Appeal that, in principle, if the Attorney-General takes no active part in the proceedings, he should neither pay nor receive costs.

CHAPTER 6 – DOMINO EFFECT

The case had an impact of momentous proportions in the Caribbean similar to a hurricane hitting each Island at the same time.

Monopoly agreements – those that were newly signed as well as those with fifteen to twenty years remaining – started to tumble not unlike a pack of dominoes standing upright next to each other after the first one has been tapped.

First it was the Eastern Caribbean States (OECS), then Jamaica, then Trinidad, then Barbados, and then eventually all the former British colonies; the monopolies started to give way to open markets, and offers for licences were being made.

As mentioned earlier, the OECS formed a governing body ECTEL to oversee Telecoms licensing, and in the governing bodies of each island, a National Telecoms Regulatory Commission (NTRC) was created as well.

Control of this lucrative industry in the OECS started to become such an issue that although the bodies were meant to be independent, the final decision to issue a licence found itself under the control of the respective government ministers, who were of course "acting upon the advice of ECTEL."

Over the next few months, in Dominica and throughout the Caribbean companies were invited to tender for fixed and mobile licences; despite the fact that there was supposed to be transparency in the licensing process, there was a cloak of secrecy shrouding the whole issue.

There were endless rumours of bribes, counter-bribes, promises of licences being issued to party A and B and C but not D, rumours of licence support for government aligned companies, and on and on it went ad infinitum.

In the meantime, the television monopoly held by Marpin Telecoms was broken when a licence was issued to another local TV company called SAT telecoms, a company that was favoured by the minister.

Some observers of the post liberalisation era described it as boys playing with men's toys.

It seemed there was no regard for the ability of small markets to sustain their companies' profitability as licenses far in excess of numbers that could survive in the market were being considered for "issue." Furthermore, it was generally felt that the deciding factors were hidden incentives to the decision makers, who themselves were hoping that licences issued would not perform well, and thus more licences could be "issued" at the benefit of decision makers.

Government ministers also appointed themselves as the ones who would hire and approve executive members of ECTEL and NTRC to ensure that "suitable" candidates were selected (and of course suitable meant favourable to their position).

CHAPTER 7 – TREACHERY AND DISLOYALTY

While all these events were taking place, the major shareholder of Marpin, Mr. John Keller, who was the main source of funding for Marpin Telecoms in the early years, began selling his shares.

As the CEO of Marpin, I made the decision to obtain a loan so that I could purchase Mr Keller's shares, but because of other financial commitments, I had to sell some of Mr Keller's shares. Mr. Clifton Shillingford bought those shares, making him the second largest shareholder of Marpin with 23% holdings.

Mr Shillingford set about to obtain control of the company with the help of some other shareholders, and he spread rumours and innuendos about me from about 1995 on, continuously causing me to spend more and more time disproving these rumours and innuendos.

Disregarding the well-being of the company and his own inability to run the company, Shillingford continued relentlessly his campaign of character assassination, even to the point of writing local banks on March 19, 2002 asking that no more loans be given to Marpin, as he claimed he was not happy with the running of the company. One bank responded on May 9, 2002 confirming compliance with his request.

The backstabbing was now having a negative impact on the company as an acute need for fibre was being felt around the island. Filling this need was seen by Marpin as a priority to maintain and strengthen our position in the market, but our efforts were slowed considerably because the banks would not provide the funds we needed, thus causing Marpin to miss out in the liberalised market. This forced us to have to do the task from our cash reserves, thus imposing even more financial hardship on the company.

SAT Telecoms saw an opportunity that it otherwise would not have had and moved in to procure, install and commission the same fibres; this gave it tremendous benefit in the new era.

Marpin's board of directors became divided over these developments, and they spent long hours discussing petty issues.

By December 2002 Marpin had secured over 1.3 million minutes of incoming telephone traffic equal to or about the same as Cable and Wireless; this was a near miraculous feat in so short a time.

There were 25,000 households in Dominica at that time, and Marpin had over 15,000 connected to cable TV. Still, the rumours and innuendos continued, having the negative effect of slowing the development of the company and putting it into ever increasing financial hardships.

Toward the end of 2002, the board voted for Mr Clifton Shillingford to be chairman of the board meetings, which was quickly followed, within two to three board meetings, with a vote for him to be chairman of the board, which coincided with the end of my contract as the CEO.

An article in the Monday, July 14, 2003 issue of the local newspaper, the *SUN*, reported these developments.

THE SUN
$1.50

Monday, July 14, 2003   Commonwealth of Dominica   Vol 5 No. XXXII

Ron vs Dense

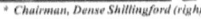

* **Chairman, Dense Shillingford (right)**

* **Chairman, Ron Abraham**

Mighty fight at MARPIN can affect future of the telecoms company

A highly acrimonious attle is developing over the ontrol of MARPIN elecoms and Broadcasting MARPIN) that threatens ie very existence of the ountry's first indigenous :lecommunications company, *The Sun* can report.

The fight, which is likely to end up in court by the time this article is published, pits the two largest shareholders, Ron Abraham and Dense Shillingford against each other and has virtually left the company with two boards of directors, sources told *The Sun*.

"We are in the middle of a battle. It's a fight for the company," one source familiar with the internal politics told *The Sun*.

An already untenable situation took a turn for the worse last week after a dramatic development in the boardroom.

Abraham, who had been demoted as chief executive officer and later sent on leave, had requested that the board of directors summon a meeting to determine if the shareholders should remove the directors, including Shillingford who is the chairman, sources told *The Sun*.

What happened after that intensified what was already a long and nasty war between Abraham, who controls a little over 38% of MARPIN shares and Shillingford, who has about 23 %, the sources said.

According to the sources, Shillingford, the chairman of the board, first sought and failed to get a court injunction to stop the meeting, then attended, and performed his function as chairman.

However, according to sources familiar with the proceedings, held last Monday night, Shillingford ended the meeting without dealing with the substantive matter, but not before Abraham had repeatedly immediately commence that meeting that was intended to be called.

"After the chairman closed the meeting, some of the shareholders left, Ron reopened the meeting to deal with the matter. The board was thrown out and a new board was elected," the sources said. "In effect there are two boards."

Cont'd on page 11

SUMMARY OF THE KEY POINTS OF THE ARTICLE

The members who were replaced were served with minutes of the meeting at which the decision would be made to fire them.

However, each board insisted it was the only legitimate one.

One source made the observation, "It is very likely that the matter will reach the court and the court will decide on the validity of Ron's meeting."

Whichever way the court would rule, the conflict was expected to continue. There was a dress rehearsal of sorts on Thursday when the police were called to increase security outside the company's offices.

Sources told the *SUN* that a member of the Shillingford board called the police to request additional security, complaining that I had entered the office and was acting in a threatening manner. The source stated that I was demanding to be allowed to return to work, and that my general demeanor was threatening.

The board member was concerned that something could have happened in light of the fact that some of the employees were agitating, and I "was in and out of the place and calling directors demanding his keys," the source added.

Two police officers were seen standing watch outside the company's offices on Thursday morning, reportedly in response to the call.

Neither Shillingford nor I would comment for this story, with both of us referring to the pending legal battle.

However, one concerned shareholder said the situation was not good for the company, which was already struggling to meet its financial commitments. "Depending on how it plays out, it could be the end of the company," stated the shareholder. "The uncertainty could make it difficult for people who are interested to invest in the company or for the banks to provide financing. It will take a while to recover."

The article continued to state that this acrimony had developed just short of the 21st anniversary of Marpin, which began operations on July 26, 1982.

With backing from businessman John Keller, who became the majority shareholder, and using funds from his other company, Maroni Electronics, Abraham began by offering four television channels.

Abraham became majority shareholder after Keller left in 1995, but he later sold about 1400 shares, which meant that a lie being circulated by another director was exposed. Also, Abraham no longer controlled over 54% of the company's shares.

Today, Marpin Telecoms and Broadcasting offers 52 channels, as well as Internet and telephone services.

However, the right to offer telephone service came with a cost from which the company is yet to recover.

It took on Cable and Wireless in a court clash that reached the Privy Council and later fought the British giant over interconnection.

The company s financial health suffered a further setback when the Eastern Caribbean Telecommunications Authority (ECTEL) refused to grant it a licence that would allow it to break into the lucrative mobile phone sector.

ECTEL ruled that Marpin was unable to raise the necessary capital to offer the service.

At a shareholders' meeting in September 2003, Mr Abraham lost his place as a director on the board. Since his position mandated that he be a board member, he was therefore forced to be out of the governance of the company even if he remained the largest shareholder.

Cellular licences were issued, and new companies AT&T and Orange were started, with AT&T sold to Cingular Wireless within a year, who in turn sold to Digicel, with Cable and Wireless in nearly total control of local market termination to a point where new companies did not even have an interconnect agreement with Marpin.

The woes of Marpin intensified as the present chairman lacked the vision and direction that were vital to grow the company to a point where it could sustain revenues and grow market shares while finding new revenue streams, not forgetting the rescheduling of debts built up during the fight with Cable and Wireless. An article in the *SUN* newspaper dated Monday 29th August 2004 again summarises the situation, not totally accurately, but enough to portray why Marpin was up for sale:

THE SUN

$1.50

Monday, August 9, 2004   Commonwealth of Dominica   Vol 6 No. XXXIV

Advertisement for the sale of Dominica's first indigenous telecommunications company appears in local and regional newspapers.

MARPIN on Sale

Dominica's first indigenous telecommunications company could end up in foreign hands.

Advertisements have started appearing in Caribbean newspapers seeking a buyer for MARPIN Telecoms & Broadcasting Company Ltd. (MARPIN).

One advertisement in the Barados Nation newspaper indicates that significant assets being old include hybrid fibre coax network throughout the island of Dominica; receiving and transmitting equipment; property, including the headquarters and premises located in a prime location in the capital city of Roseau"; fleet of motor vehicles and other assets including spares, tools and other support equipment.

"All I have done is put an advertisement inviting offers for purchase," the receiver, Michael Toney told *The Sun* in a telephone interview from his office in Port of Spain, Trinidad.

"We hope we can get serious bids for the offer of purchase so the company can continue operating under new management and give everybody a chance to breath a little easier," he added.

Toney would neither indicate what he thought a reasonable offer would be saying that giving out this information "will affect the fairness of the process" nor would he disclose the level of the company's debt, telling *The Sun*: "The debt cannot be disclosed because that is private information."

However, in a February 24, 2004 report to "receivers and others", the first receiver/manager, Gordon Moreau, indicated that the company's net assets, including cash, account receivables and inventories at EC$1,071,505.00 and current liabilities at EC$26,420,596.00. The net liabilities were put at 11,331,038.00.

The company's value is estimated to be in the region of EC$40 million according to a source familiar MARPIN, although another source with previous information on the workings of the company put the value at somewhere between EC$5 million and EC$10 million.

Persons associated with the company worry that in his "rush

to sell the company and get out of here" Toney would allow MARPIN to go for something in the region of the EC$10 million which the second source estimated that it is worth. This would mean, according to one concerned individual, that unsecured creditors and shareholders could be left hanging with nothing to show for their investments.

"The only obligation is to pay government, then the receiver, then the workers, then the bank (National Bank of Dominica), then you put it together you can then bid," one source observed. "Somebody can get it for twelve million (dollars). The (unsecured) creditors could lose everything, the shareholders will lose every-

thing."

Bids close at 4:30 PM on October 4, 2004 and advertisement appearing outside of Dominica there is concern that the company could end up in the hands of non Dominicans.

"It would be of concern to u particularly if that company i known to be anti worker in thei approach," Curtis Augustus of the Waterfront and Allied Worker Union (WAWU), the union representing MARPIN employees told *The Sun*.

"We would prefer if it could be retained in local hands... If it happens to go into the hands of a foreign entity we would prefer if this foreign entity would be prepared to come to Dominica and respect the workers and not violate the labour laws and cause (damage to the industrial relations environment," the union boss added.

While Toney told *The Sun* that there was no guarantee that a buyer would be prepared to retain the staff, Augustus said he was not overly worried at this time about staff tenure since the labour laws made provisions for staff to be retained or severed and rehired.

"When we last met (Toney) in dicated that if there were any developments, positive or negative he would get back to us, Augustus stated. "We are waiting and the fact that he hasn't, we are looking at it optimistically."

Meantime, a number of unsecured creditors are prepared to help save MARPIN to avoid "losing everything," *The Sun* has been told.

"The creditors will lose everything if it (the company) is sold and if they can get 25 cents on every dollar they would prefer that than to lose it all," one source said. One of the main creditors behind the move is MCI, America's second largest long distance carrier which MARPIN owes over EC$9 million, according to one source MCI has already contacted Toney with a proposal and officials of the US company are likely to visit Dominica "to put it together," the source said.

Toward the end of 2003, the creditors became concerned that the company would go bankrupt, and therefore moved in through the courts to secure their interests. National Bank of Dominica paid off the other secured creditor, First Caribbean Bank, and appointed a receiver-manager in the name of Gordon Moreau. He was replaced by Price Waterhouse, who was then replaced by Anthony Toney from Trinidad.

These actions in effect removed control from the board and from the chairman. So, after all the plotting and scheming to remove me as chairman and CEO, it became an exercise in futility, as they were only in control of the company for one year.

CHAPTER 8 – LESSONS LEARNED

No matter what you experience in life, every encounter, every chance meeting, every struggle, and every hardship holds within it lessons to be learned. Later in life when you look back on these struggles and wonder why they had to occur, you will often discover that you've gleaned valuable insights that will stay with you the rest of your life.

In this David-and-Goliath struggle of Marpin versus Cable and Wireless, however, many lessons were learned. In order to bring them into perspective, several questions need to be considered.

QUESTION:

Concerning the pressures that were placed on me to sell the shares I had purchased (as CEO) from John Keller; was that orchestrated as part of a sinister plan to remove me?

ANSWER:

> The manager of the bank at the time was a good friend of Mr Shillingford, who held less than 1% of Marpin's shares. Could it be possible that having heard through the grapevine that I had purchased Keller's shares, he decided that a friendly request could put pressure on me and force me to sell some shares to Shillingford? In my opinion, it is very likely this was what happened, but it will have to be left to the imagination, as we'll never know.

QUESTION:

Were significant errors made in a count of ballots at an annual shareholders' meeting in 2003, resulting in then deputy Chairman Clifton Shillingford remaining on the Board of Directors? Furthermore, was this part of the same sinister plan to remove me?

ANSWER:

In 2002, sensing the move afoot to remove me by Mr Shillingford, I obtained proxies that give me about 49% of votes at the annual shareholders' meeting. When the votes for Mr Shillingford were counted, the representative of an accounting firm determined that Shillingford had won 9,000 votes out of 11,000. I challenged the validity of this count, as the meeting minutes reflect. The votes were recounted by the same person, who came up with 5,800 votes for Shillingford on the second count. In a bid to avoid controversy, I did not object at the time. However, there was a serious error in law; had the ballots been rechecked after the meeting, it would have been discovered that the true count was only 4,800 votes, meaning that Mr Shillingford would have been removed from the board. On checking with legal counsel, it was determined that the declared result had to stand. Investigations revealed that the person doing the count was the sister-in-law of Mr Shillingford, something I had been completely unaware of. I will let the reader determine the validity of a family member being the vote-counter!

QUESTION:

Would it have been better for Marpin to strike a deal with Cable and Wireless to obtain a duopoly instead of fighting to break the monopoly?

ANSWER:

On hindsight, it probably would have been better for Marpin to be a reseller for Cable and Wireless and expand into telephony in collaboration with Cable and Wireless. That would have given Marpin protection for a long time into the future, and would have saved the high cost of the battle and all the backbiting and jealousy that I had to endure in the aftermath of the liberalisation. It would not have been beneficial for the Caribbean, but the company would have been better off. Did Marpin follow the best course for Dominica? I certainly think so.

QUESTION:

Should the Caribbean governments have come together to support Marpin in light of the significant benefits obtained by them because of the battle?

ANSWER:

The governments of the Caribbean all benefited greatly from the battle that Marpin fought to break the telecoms monopoly, and in my opinion, they should have ensured that Marpin received the same licences and privileges as Cable and Wireless to support our growth as the only local telecoms company in the Caribbean at that time.

QUESTION:

Was the refusal to grandfather my company, Marpin, into the field of cellular telephone service providers also part of the same sinister plan to remove me as CEO of Marpin?

ANSWER:

In looking back at the situation after the passage of time, it seems plausible that some sort of plot was hatched from within Marpin and in Dominica to put me down, but it is all a matter of conjecture.

QUESTION:

Would a favourable vote (toward me) by the Company Secretary Burnette-Biscombe at a crucial board meeting when Clifton Dense Shillingford was voted as Chairman of the Board meetings have averted the post liberalisation crisis of Marpin? Was Burnette-Biscombe given a carrot?

ANSWER:

At a board meeting, when Shillingford motioned to have me removed as Chairman and replaced with him, the voting was three for Shillingford and two for me. The last to vote was the company Secretary; he knew what he was going to do beforehand and apologised to me for having voted against me, thus making it four votes for Shillingford and two for me. Had he voted for Abraham, there would have been a three-to-three tie, which would have been tipped in my favour with deciding vote as Chairman. Could it be that a carrot was offered? Again, it is a matter of conjecture.

QUESTION:

Would Marpin have been better served if I had been more involved in the day-to-day political events in Dominica?

ANSWER:

Small countries can be very stressful since everyone knows everyone and sensitivities can very easily be aroused. As an example, if your mind is occupied with pressing thoughts and you pass someone on the street and do not notice them or acknowledge them, you have possibly made an enemy. Likewise, because you get invited to a countless stream of cocktail parties, sometimes you are simply unable to attend one, but in choosing not to attend, you have just made an enemy. These sorts of misunderstandings make it difficult to conduct business dealings in a small island nation.

CLOSING THOUGHTS

Trying to make a company successful in an island nation (and region) that is short of so many skilled citizens, one often has to do many different jobs towards making things work requiring long hours, sometimes working seven days a week. This leaves one unresponsive to situations and issues that seem petty at first glance, but in a small country, where everyone knows everyone else, nothing is ever petty. Even issues that begin small can grow in significance and seeming importance with each bit of shared gossip and each retelling of the story, and those small issues soon have begun too large to ignore … and sometimes that realisation comes too late.

As the main driving force of Marpin, I as the CEO probably should have spent a lot more time socialising among key people in Dominica (and elsewhere) to have friends in my corner when the going got tough. The successful rate of growth of Marpin possibly would have been smaller as a result, but maybe the company would have survived.

In conclusion, it was a life shaking experience to tackle Cable and Wireless and bring down their monopoly, and to see the vast benefits to the region as a result, but at what a steep price victory can be won and lost … what a price indeed!

In all these things, we give thanks to the Lord Jesus for being with Marpin throughout and for empowering me, His servant, with wisdom, grace and fortitude.

ABOUT THE AUTHOR

The author was born in Dominica, then a British colony.

At age 20 he migrated to UK where he stayed continuously for over 20 years. While there he studied Electronics, Electrical Technology and Telecommunication principles as well as experience having worked with Graviner and Cossor Electronics now Ratheon.

He was driven to go back to the Caribbean with intention to contribute something of substance to the region.

About 1976 he worked with Guyana Telecoms as a Executive Engineer in Guyana before returning to Dominica in 1979.

He set up a calibration facility but still driven by a desire to do something meaningful he closed it up to open a Television Company Marpin Telecoms.

This book is about the struggle including the breaking of the Telecoms monopoly of a multinational successfully.

While doing that forces of Jealousy and greed drove him out of the company and reflections on whether it could have ended differently.